I0035997

Balance Theory

Parviz Razban Haghighi

2013 © Parviz Razban Haghighi

All Right Reserved

ISBN: 978-1939123176

Publisher: Supreme Century, USA

BALANCE THEORY

Introduction

In spite of all progresses of science in recognition of existence nature, still it is not able to answer so many question.
Like: what is big bang nature and how this world has been created from nothing? Or what is the nature of light and what are moves and in spite of its intensity and weakness, its speed is equal in all directions? Or what is the essence of time and how it has been created? Or other questions regarding space, substance and others....
As per my researches in this regard, I realized that with the current level of science we are having today, we are not able to provide answer to these questions.
Therefore, after performing studies and researches, I reached new and at the same time fascinating results and finding, which I will present them in terms of Balance Theory which I hope will be a step toward a better understanding and recognition of the world.

Parviz Razban h.

parvizraz@gmail.com

For understanding the nature of the world, as the first step we should get familiar with new mathematics which I am calling them curve mathematics or balance mathematics, which are completely different with the mathematics we used to know so far.

The mathematics we used to know so far and I call it linear mathematics cannot be of so much of help for us in knowing the nature and essence of the world and is not capable of providing us with answers in this regard. off course this my sound a bit strange, as it is thousands of years that human are applying and using this type of mathematics and the logic of human are based on it. But the reality is that, this type of mathematics is shaped based on the world we are in and only has application based on the existence of this world. But regarding, the non-existence and existence and their relationship with each other, is not able to make the sense it should make. For this reason only I have invented a new type of mathematics. This mathematics is more consistent with the reality engulfing the existence and non-existence and in a better way explains the logic encompassing it.

BALANCE MATHEMATICS

Contrary to linear mathematics in which numbers grow in line with each other and with equal distance from each other. In balance mathematics the growth of numbers have a curve shape. As much as the number is bigger, its distance becomes smaller with the next number. For a better understanding you should see the following figure:

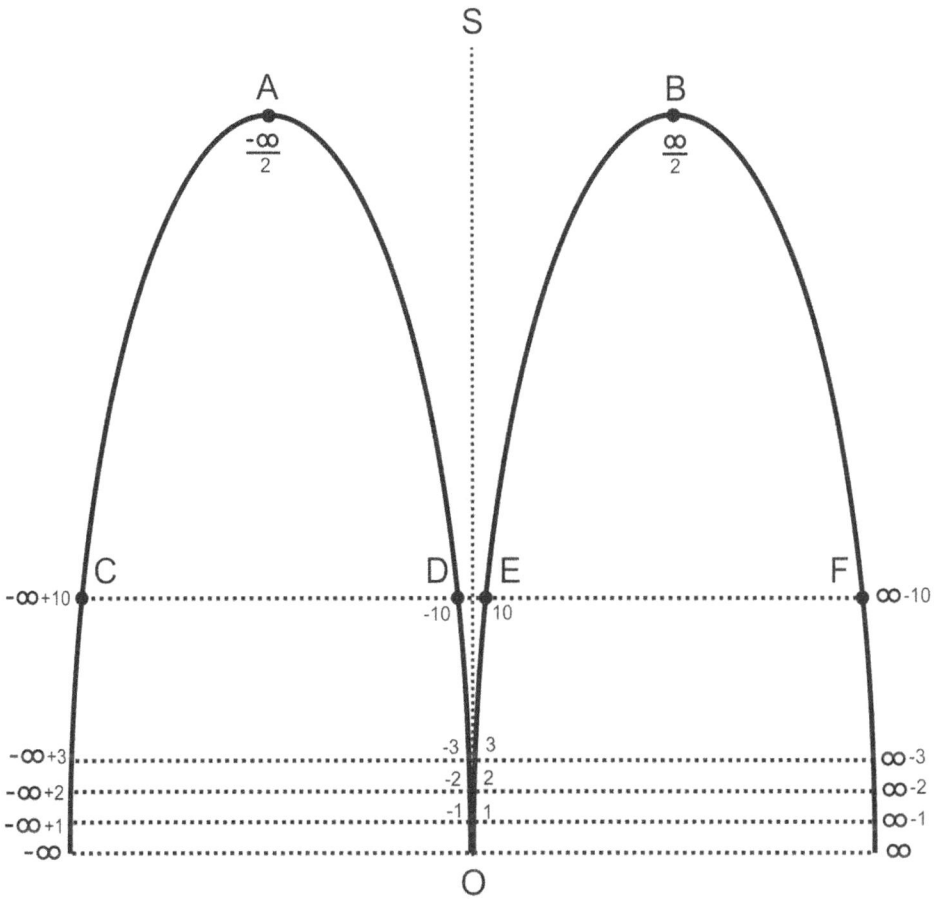

figure 1

Numbers in balance mathematics, increase in a curve shape from.

Each number has a specific amount of ambiguity or load and as much as the number is bigger, this ambiguity or load is bigger as well.

Zero is equal to infinite or in other words they are of the same value.

As you can see in figure 1, the sequence of numbers is seen in two curves. On of the curves is the curve of positive numbers and the other one is the curve of negative numbers. Each number is equal to its negative one. For example, consider number 10 is equal to -10.

Contrary to linear mathematics in which the axis of 10 and -10 is in line with the axis of S series of numbers. In balance mathematics the axis of S is perpendicular to the axis of 10 and -10 (figures 2 & 3).

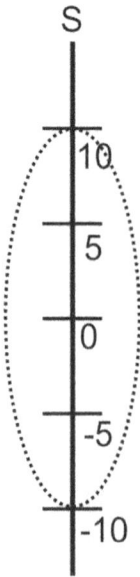

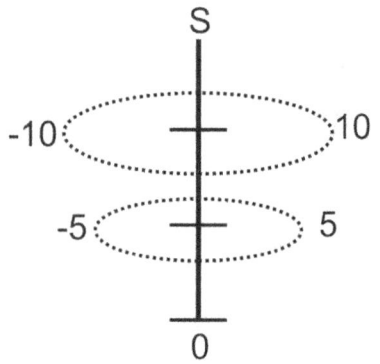

Figure 2   Figure 3

As it is seen in figure 1, number 10 has crossed these two curves in 4 location. The distance of 10 with -10 or the DE segment is considered the load or the ambiguity of the number and is calculated as per the following formula:

$$W = \frac{2X}{\infty}$$ W= load or ambiguity   X= number

It should be noted that in balance mathematics, the number divided by infinite is not equal to zero and the amount of its curve is the number. It is because each number in the curve of the series of numbers, has a specific location and partnership, therefore infinite divided on a number will not become equal to infinite, it will become a specific number. In figure 1, number 10 in addition to points of DE, has crossed the two location of CF as well. For knowing these two points, it is better to know two points of A and B. These two points is the location of number $\frac{-\infty}{2}$ and $\frac{\infty}{2}$ and is the turning point of the numbers. In balance mathematics, as the numbers increase, their distance becomes smaller or in fact some of the distance is transformed into the load of the number. In points of A and B, all the number distance has transformed into load.

$$W = \frac{2\,\frac{\infty}{2}}{\infty} = 1$$

Therefore, after the number $\frac{\infty}{2}$ with increasing, the number is descending instead of ascending or other words, its load will be bigger than its number distance and will cause a decrease in its numerical position.

$$\frac{\infty}{2} + 1 = \frac{\infty}{2} - 1$$

The inverse of this action is happened at the location of A. Hence, in CF points in figure 1 we will have equal numbers of 10 and -10 but in descending manner.

Four points of CDEF in fact from a number. Hence, each number has two personalities. One of them has heavy weight or ambiguous character of CF and other one has light weight or unambiguous character of DE.

Each of these characters has two dimensions, ascending and descending. The ambiguous character is the inverse of unambiguous character. Then, the inverse of each unambiguous number is ambiguous and unrecognizable. It is because its ambiguity value is more than its numerical identifier.

In the light weight character of each number, as much as the distance between the two poles of the number is smaller, the number will have a lower weight or ambiguity. The number of zero, which its two poles are on each other, has zero amount of ambiguity or has a complete certainty and its inverse ∞ & $-\infty$ possesses the highest ambiguity.

The number of $\frac{\infty}{2}$ and $-\frac{\infty}{2}$ is exactly on its inverse and therefore we cannot recognize it from its inverse value, it is neither ascending nor descending. This number can be called a neutral zero or real zero due to its characteristics.

The sum of each number with its inverse is zero or infinite and the average of each number with its inverse is neutral zero.

Each math operation done between numbers, automatically the negative of it (the operation) is performed at its negative part and therefore would neutralize result of the operation. For example:

$5 * 4 = 20$ in positive part

$-5 ⊛ -4 = -20$ in negative part

⊛ is negative of multiplying

In inverse section of numbers, the inverse of these two operations will occur and would neutralize the clarity of these two operations.

In balance geometry of each geometrical shape depends on the position of the observer and is seen in a specific way and with changing the position of the observer the geometrical shape is seen another shape, and also the negative geometrical shape exists in the negative part. The inverse from of these shapes is created in the inverse part in an ambiguous form.

No geometrical shape will exist in infinite dimension.

It is possible that this question will be raised that why the existence and non-existence relationship of linear mathematics are not so much responsive. In response to this question we should say that this mathematics only has relationship with existence and explains the logic ruling it. However, it faces problem when it relates to with the whole reality which includes the existence and non-existence worlds. Regarding the whole reality, the mathematics science has emerged from nothing, so all its numbers and

All its operations in all the fields should be neutral zero in the sum. For example in:

$2 * 5 = 10$

for the whole reality it is not logical why 2 or 5 should reach the position of 10. Why the operation of ascending takes place and from where it comes.

This operation is contrary to mathematical balance, while in the negative part of numbers, the negative multiply is performed and decreases the number of -2 or -5 and brings them to -10 and hence the balance is maintained. In inverse part of the numbers, the inverse ambiguous of this operation is performed and neutralizes its clarity.

In drawing a geometrical shape, with automatic drawing of it in the negative part and by drawing its ambiguous part automatically in the inverse part, either in the positive inverse part and in negative inverse part the balance is maintained and in general would make the shape neutral zero.

Contrary to linear mathematics which has inertia, the balance mathematics has movement. If we imagine a number, its position in linear mathematics is always fixed and no changes happen to it. But in balance mathematics as soon as we imagine it due to its characteristics, it will start to grow. For example we imagine number 5, its weight is:

$$W = \frac{10}{\infty}$$

But number 10 is also having a weight. therefore:

$$W = \cfrac{10 + \cfrac{20}{\infty}}{\infty}$$

Also the weight of number 20 is added to it and this will go on till infinite. Therefore immediately after imagining the number 5, this number is coming out of its position and starts to grow. Then it will only have the value of 5, at a cross section of time.

Hence, we can say that balance mathematics has a virtual timing. In inverse part of numbers the inverse of this action occurs. Then we can say that it is having an inverse virtual time.

Neutral zero or real zero is the gravity point of numbers. It is because it absorbs numbers to itself after zero and in inverse section before infinite. The inertia locations of numbers are zero-infinite and neutral zero.

BALANCE THEORY PHYSIC

This theory is based on two principles:
1- The existence world and all its shape is formed out of photon.
2- Photon has two contradictory and binary nature, called existence photon and non-existence photon.

PRIMARY WORLD OR IDEAL WORLD

As it was said photon has two contradictory characteristics. In this section some explanation will be provided in this regard.

1st circumstance: photon doesn't have any dimension and it doesn't have any time and has infinite density. In this case it is called non-existence photon.

2nd circumstance: photon has infinite dimensions and infinite fast time and its density is zero. In this case it is called existence photon. Existence photon is the inverse of non-existence photon.

In this circumstance the two photons have the most difference with each other. Although both of them belongs to photon character, but for a better understanding we will consider them apart and independent from each other.

In this case due to the phase difference which exist between existence and non-existence, the non-existence photon will move with infinite speed inside the existence photon and since it is not able to do anything about this phase difference

Therefore this movement will continue always and in all the directions.

I have named this situation primary world or ideal world. This world is not visible or it can be said that it doesn't exist in the real world. The ideal world is the foundation of the visible world and is absolute. It always exists and all the shapes of world are only definable and recognizable based on it.

The ideal world in fact is a kind of logical imagination.

It is possible that we will have this question that why the existence and non-existence world or ideal world should exist?

In response to this question we should say that if we will be able to separate all the elements of world including particles, energy and space from the world, all that remains is non-existence and separating it, is like recreating it and it will never disappear. The non-existence which is made of separating existence from it, is the certain non-existence and only exist opposite to existence and existence should be present compulsorily in order to certain non-existence to find meaning. In fact, certain non-existence is equal to zero and existence is infinite or the inverse of non-existence. Therefore, the ideal world will always do exist.

PHOTONAL

As it was mentioned non-existence photon is moving with infinite speed inside the existence photon. As this movement is in all the direction and in all the existence points and

Locations, in some locations some nodes are created or in other words the photon is caught captive in its own trap.

I name this node as photonal.

photonal is the existence crest and non-existence inside. From the point of view of non-existence inside, photonal is without dimensions, without time, with density of infinite and without movement.

From the point of view of existence crest, photonal have infinite dimensions, infinite time and density equal to zero and its non-existence inside (or core) is going around itself with infinite speed.

Photonal is at the emergence threshold in the real world or in other worlds, is at creation threshold of real world.

These photonals can make connection with each other. With the first connection made between the photonals, the world is emerged or created for them. This moment is called the moment of big bang and the time is started for them from this point.

Off course in the condition that photonals are, quantity doesn't have meaning and connection of them is not so much correct. The more correct is to say that the existence crest of photonal penetrates into its non-existence and on the other hand non-existence also is injected into the inside of existence.

The inside of non-existence is transformed in to the real world and the crest of existence makes or creates the inverse world.

In reality world with connection of photonals to each other,

The connection between them which is from the existence material, would decrease the density. As much as the density decreases, the world becomes bigger for them. In the inverse world the contrary to it occurs. The density increases and the world becomes smaller.

In the real world and the inverse world, zero and infinite doesn't exist.

Photonals get combined with each other and create black string objects, which in turn creates all the substances and objects in the world.

INVERSE WORLD

From logical point of view the existence of inverse world is necessary. It is because it will give balance to the real world and general would make all the particles and objects and the interactions between them neutral zero and hence no contradiction is created with the ideal world or before the big bang moment. All the particles as they are existing in the real world, exist in the inverse world as well, but due to high amount of ambiguity of the inverse world, they only feel their existence in the real world.

In addition to neutralizing the real world, the inverse world would determine the clarity amount of each particle in the real world.

Each particle has an existence coefficient. As much as the existence penetration would be more from the inverse world to the inside of the particle, it will have a higher existence

Coefficient. As it was mentioned in photonal section, the connection between the photonals is of existence material. In fact all the connections between the particles are of existence materials. As much as the particles are bigger, more existence is needed for the connection between them. Foe example, we need more existence for connection between the elements of atoms including protons and neutrons, comparing for the connection between the photonals. As a result, atom has a considerably less density and more existence comparing to black string objects. Hence, the existence coefficient of atom is higher.

So in this way, a molecule has a higher existence coefficient comparing to atom and similarly compound substances has a higher existence coefficient comparing to a molecule.

As much as the existence coefficient of a particle is smaller, it has more phase difference comparing to its inverse. As much as the phase difference is bigger, the particle will be more clear and its inverse is more ambiguous.

If we consider a particle as a galaxy, it has a very higher existence coefficient comparing to an atom. As a result, it will have more ambiguity in dimensions, movement and time comparing to an atom.

Regarding the whole of the real world, we should say that it is coincident on its inverse and completely locks clarity. Or in other words we can say that it is neutralized zero. Now we will face this question that where is the inverse world and what is its nature and essence?

According to what has been said previously, each particle has an inverse.

The particle itself has an infrastructure of non-existence and its inverse has the infrastructure of existence. But in response to the question that where is the inverse of the particle? We should say that for better understanding and recognition of the location of the particle it is important to know where it is located and it is equally important to know where it is not located. In fact where the particle is not located, is the part of the inverse particle. In the next topic it will be realized that each particle has exclusive space. Therefore, the space of each particle is the inverse section of the every particle.

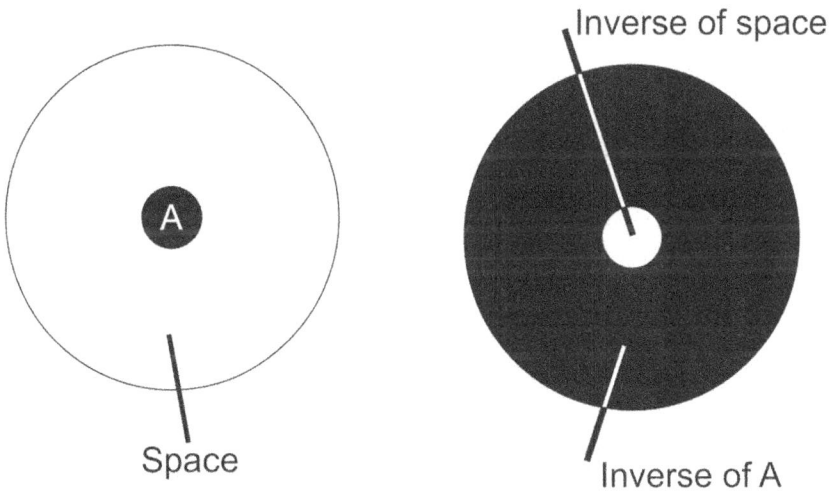

In the inverse section of the particle, all the particles and objects which exist in this world and all their interactions are observed in an inverse from.

For example, if the light wave will travel from point A to point B, the observer (in inverse) will feel that section of space which doesn't have wave. Therefore it will feel that the wave has gone from B to A and since time is created by the movement of photon, therefore, time would be in an inverse from in the inverse world, however it doesn't mean that time will turn back in time and only the position of past and future is changes and it is so much different than the fact the future and past will be in their own position and place and the time will go back to past.

From the point of view of speed the time will be as following:

$$ST' = \infty - ST$$

ST is the time speed of the particle and ST' is the inverse time of the particle.

In contrary to the real world, if the distance between two particles is from outside, in the inverse section, the observer of the particle, would feel this particle inside itself.

For example, if we would assume you to be a particle, all the real world around including the objects surrounding and all the galaxies and planets and .... In the inverse world will be placed inside you.

In the previous section it was mentioned that all substances and different shapes them which exist in the real world, have the root of non-existence photon or in the other worlds, the smallest particle making an object is non-existence photon.

Therefore, the smallest particle making you is non-existence photon with zero dimensions and zero time. The inverse of non-existence photon has the highest dimensions, infinite and has the highest speed of time, infinite speed.

Therefore, in inverse section, you are positioned inside yourself in an inverse form and have inverse time. You see the real world from outside which is growing with the speed of C, while in your inverse section, this world from inside is becoming smaller at the speed of C.

Since the real world has root in non-existence photon, it is clear and understandable. But the inverse world, since it has its root in existence photon, is ambiguous and without clarity. As much as the existence coefficient of a particle is lower, its inverse will be more ambiguous.

Perhaps it will seen strange but the presence of inverse world is necessary beside the real world. It is because that in all the cases either regarding the single particles or regarding the whole particles of the real world and their connection and interactions with each other and in all the moments in general will put the circumstances in the condition of neutral zero and keeps the balance at the condition of ideal world.

SPACE

As it was mentioned earlier, existence photon has infinite dimensions and non-existence photon has zero dimensions. Each particle is real world has an infrastructure of non-existence inside of photonal that existence have penetrated inside its crest. Therefore, each particle has perspective of existence equal to the amount of penetrated existence to its inside which is called space.

The characteristics of space is as per the following:

1- space is extension of C. (C is the speed of non-existence photon)
2- space is the very same time. Contrary to common belief that consider time as the 4th dimension of space, in fact the 3 dimension of space are the dimensions of time also.
3- each particle has an exclusive space which is an integral part of the particle and is one of the main elements of the particle.
4- each particle is located at the center of its space and is not able move in its space.
5- the space dimensions of each particle depend on the existence coefficient of the particle and as much as the existence coefficient of the particle is larger, it has a bigger or more expanded space.
6- due to high amount of ambiguity existing in the space, no particle is able perceive the dimensions of its space without the help of other particles.

7- each particle is present in the space of other particles without its space and other particles without their space are present at the space of the particle.

8- the space of all the particles are compound and it is from the combination of the primary space which is created of the connection of two photonals.

9- each particle has special perspective toward other particles due to having an exclusive space.

Due to characteristics of space which was explained earlier, it is concluded that each consequence which is happened to the particle, all the particles in it space are in a special way affected by it which is different comparing to the effect of it in the spaces of other particles.

For example, if a star under the effect of supernova, changes into a neutron star, all the particles in its space will change. For example, the planet earth will become closer to it and density will increase. While, in the space of the planet earth, that star will only give more curve to the space.

As it has been mentioned, space is the very same time. It means that all their characteristics should be true about both of them. Time has future, present and past. Therefore, space should posses the same characteristics. In this regard we should say that the photon coming near the particle is future, the one going far from it is the past and the one collided with the particle is the present of the particle. The present period of time of a particle is the time that photon is passing through it.

The light which comes from a star to us is in our future and the light which goes from us toward that star is in our past and the light reached from that star to our eye is the present time. In fact, our present time is a period in which photon passes through our body.

However we should notice that photon always exists and space is created from its movement and it is not that it only exists in light or electromagnetic waves. Light has a totally different nature. Light is wave. Its only difference with the water wave is that it rides on photon.

As it was mentioned no particle is able to move in its own space. The movement of each particle is occurred in the space of other particles.

When two particles are inertia to each other, it means that they are balance from the point of view of resultant forces. Now we if apply a force to one of them which would send them out of balance, each of the particles would start to move in the space of the other. But for knowing which of the particles is moving, it is better to say that the particle that its existence coefficient has decreased or the one with a decreased space and slower time.

As much as the existence coefficient of a particle would decrease, its speed would increase. Only non-existence photon can reach the speed of C, because it is possessing the existence coefficient of zero and has the highest contradiction with the existence of the space of other particles.

Other particles cannot reach such contradiction due to the penetration of existence into them. As a result they will never get the speed of C. The movement is always opposite to the movement of time and expansion of the space.

TIME

It seems one of the most unknown forms of the real world is time. First we should see what is time?
As it was described earlier in the section of space, time is the same as space. If we want to state it better we can say that we see time in the form of space and we feel space in the form of time. Now we want to discuss this matter in a more clear way.
What is the essence of time? As it was described in the section of photonal, the crest of existence, sees the inside of its non-existence turning around itself with the speed of infinite. Now with injection existence into non-existence, it will also feel inside the turning around equal to the penetration of existence. This turning is time. This turning created the existence perspective or space for the particle.

Time has two elements:
1- The element of time that everybody are familiar with it and is shown with T which indicate the passage of time and the interaction of the particles are performed in this section.
2- The element of existence coefficient which is in vertical position on T and is totally unknown so far and is shown with Ts.

Consider the following figure:

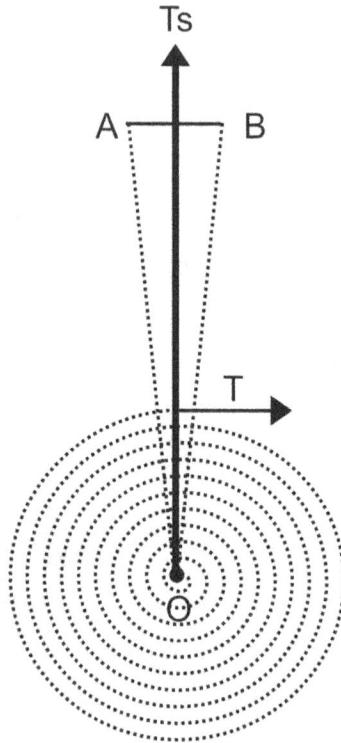

Figure of time

T is the movement of time. This movement is not consistent and depends on the position of particle and the effect other particles have on it which will change it as well.

Ts in fact is the present time of the particle that the particle is on it always and is as well the existence coefficient of the particle.

As much as the existence coefficient if a particle is bigger,
it has a bigger present time.
AB segment indicate the size of existence coefficient or the
present time.
The movement rhythm of Ts on T in all the particles is similar
but due to its size it, can have a faster or slower T. Exactly
like that all have equal numbers of steps but some have
bigger and some have smaller ones.

The point of O is zero time or photonal time and is the
big bang moment. Big bang is not only the starting moment
of T but also Ts starts in this moment. Off course we should
note that all the occurrences of the real world only occurs
in the element of T, but since in big bang time T is coincident
on Ts, exceptionally this event is in Ts and is the connection
point of all particles.
Although all the particles have the big bang moment in their
present time, but have distance with it equal to the their
existence coefficient.
As it was described in photonal topic, in effect of
combination of photonals or penetration of existence into
non-existence of photonals, big bang moment is started and
time starts to move. Due to the constant penetration of
existence into the particle after the big bang, the size of Ts
increases moment by moment and the speed of T increases
as well. This action manifests the expand with speed of C in
space.

Returning back to past in element of T is impossible due to expansion. Because this expansion would only stops when the particle changes into photon. Even if we assume that such a thing would happen, the time would stop and will not return back to past.

But in Ts element returning back to past is possible, such as when a particle is moving or when its speed is increasing or when it is collapsing (like in the center of supernova), it causes the existence coefficient of the particle to decrease or make the Ts become smaller and as a result its distance would decrease with the big bang moment and then the Ts of particle is moved to past and T would become slower.

For abetter understanding of Ts and T we would assume them to be clock needle and the movement of clock needle. Respectively, at the center of the clock and the location of its attachment to the needle there is no movement. So it is considered as the big bang moment. As we go nearer to the tip of the needle, or the existence coefficient of the particle increases, the movement of T become faster.

The dimensions of the real world is from zero to $\frac{\infty}{2}$ and the dimensions of the inverse world is from $\frac{\infty}{2}$ to ∞ . As space is the same as time, then these dimensions should be observed in time also. The time from the big bang moment to $\frac{\infty}{2}$ should exist in the real world. But how is it possible that at same time all of these moment would exist.

The answer to this question is available in the structure and characteristics of the particles.

For example for us who are having a molecular structure, around 14 billion years have been spend from the big bang moment. However, for the atoms inside our body, which are having a smaller existence coefficient, due to the slow passage of time, less time has spent from the big bang moment and in turn, this time would be even lower or lesser for protons and neutrons. For black string objects inside us, due to the fact that passage of time is so much slow, they feel themselves so much near to big bang moment and for our inside photonals the moment is equal to big bang moment. For galaxies due to their higher existence coefficient, big bang moment has occurred so much time ago.

Regarding the T element, we can say, for instance, it is having a more past dense proton and a more past dilute galaxy.

If regarding the time, we only have belief in T element, how we can explain those satellites whose clocks works slower than the earth clock and hence has in a more past time comparing to earth and how is possible that they are existing at our present time.

It is possible that this question would occur to us, if the possibility of returning back in time to past in Ts element, is it possible that this event would have happened for the whole world and the real world collapse and the occurrence of reverse big bang would ruin the whole world?

In response to this question we should say, that this is an inclusive feeling for each particle.

For example, if we accept that due to high gravity in the center of black hole, photon is not able to escape, as a result of this we should accept that the mentioned center has turned into photonal or as a matter of fact the mentioned center doesn't exist, and the high gravity as a cause of the surrounding of the black hole which still exists, is applied to the photon and traps it in its center.

From these we can conclude that the center of the black hole has experienced the complete collapse of the world and in fact all the particles of the universe including the earth and us and humans, have been taking part in that collapse. So how come that we as humans, haven't felt anything?

In response to this question we should say that as it was explained in the section of space, each particle without its space, is participating in another particle's space, so we also have participated in the collapse without our own space and with our own space we will not feel such a feeling. It is because, all the time, we are inside the photonal with our own space.

DISTANCE

At the time of discussing the space it was said that the observer on particle is not able to perceive the dimensions of its space without the help of other particles. This means that space alone cannot have a scale for measuring dimensions.

Therefore, what specifies the distance of a particle from another particle in its space or in general what is the concept of distance at all?

As it was mentioned earlier, the movement rhythm of Ts on T is equal and in all particles. Now if we mark a point of T of particle, the T of all the particles in the space of that particles will marked. This point is named Dts. These points are not in one line.

If we draw a vertical line from Dts of the intended particle on the element of T, the difference of the Dts of other particles with this line will be seen as distance. As much as this difference is bigger, that particle will have more or bigger distance comparing to the intended particle in its space.

In fact if we want that the intended particle would put other particles in its own present time and would be observable foe it, it has to have such distance with them.

For getting near or far from a particle, the particle has to decrease its existence coefficient at the direction of that particle and this is done through movement.

THE REAL WORLD

The real world can have dimensions of $\frac{\infty}{2}$ and a time at the speed of $\frac{\infty}{2}$

It means that the space of each particle can grow to this amount and after that it will destroy and will transform to neutral zero.

At this moment the particle to some extent will lose its density and will become one with its space and its time will get so much speed that it will not be able to operate in that speed and it is because it will be coincident on its inverse and the observer of the particle will not recognize it and the particle will come out of the real world and will become absolute non-existent.

Neutral zero is the gravity point of the real world and this gravity is applied on particle by means of time. This gravity is the inverse gravity existing between particles. It is because particle gravity decreases the existence coefficient of particle and neutral zero gravity increases the existence coefficient of the particle. But how is this gravity applied by means of time.

In relativity physics, the particle movement will slow the time of particle and make the dimensions of the particle smaller. This matter is expressed in balance theory in form of decreasing the existence coefficient of the particle and returning to the past in the element of Ts. The inverse of this action is also true. It means with the passage of time the dimensions of the particle increases and the density of it decreases and its time becomes faster.

Since time is the very same space, then the space of the particle increases in size.
This issue is manifested by space growth with the speed of C.

The growth of space is due to constant penetration of existence from big bang point with the use of the inverse of the particle. Existence penetration will increase the dimensions of the particle and decreases its density and as a result the perspective of existence of the particle, space, is increases.

This action would result in distancing the particles from each other in the space of the particle and in general would increase the dilutions of the particle space.

Due to constant increase in speed of time, the distancing of particles from each other is increased constantly. As a result the explosion of big bang not only doesn't become slower gradually but it becomes faster as well.

Since for us approximately 14 billion years from big bang has been spent, as per the descriptions of the distance section, big bang should be at our 14 billion light years distance of us. Its means that constantly existence is entering to our space from there.

It is possible to face this question that if big bang always exists and the inverse of particle, is pouring existence from that point to the inside of particle, is it possible that a particle is born before 14 billion years?
In response we should say that for the location in which we are located, such a thing is not possible. It is because there is no time existing before it, but as it was described in the topic of time, for particles with higher existence coefficient,

Due to the higher speed of time for them, they feel themselves mor far from the big bang comparing to us.

Although certain zero and neutral zero in balance mathematics has a completely different essence but since it doesn't exist in reality, are similar for the particle.

For example, big bang is located at distance of 14 billion light years. As per the definition presented in previous section, the photon which comes near us, is in future and the one getting far from us is our past. As a result, the photon coming from big bang moment toward us is our future and the one which is going from our side to it, is our past. Then eventually past and future of us will be coincident on each other. In fact in time $\frac{\infty}{2}$ the particle not only will reach to the end of its time, but at the same time will reach to the starting point of itself big bang.

With passage of time, the existence coefficient of the particle increases and as a result, ambiguity will increases in the particle and its space. Before explaining this, it is better that first we will see what the meaning of ambiguity is.

In the section of inverse world it was said that galaxy has more ambiguity comparing to atom. Which is due to a higher existence coefficient.
You are never able to know the exact space position of galaxy accurately. It is because if we assume that the galaxy diameter is 100000 light years and you are standing at the

Corner of the galaxy, for observing the end position of the galaxy you have to wait for 100000 years. and till that time for sure the position of it have changed a lot.

And for example if the observer of the whole galaxy wants to notice that it has collided with another galaxy, he needs 100000 years. This amount of time is the present time of the galaxy. While atom clear in these cases.

So as much as the dimensions of the particle are increasing, its clarity decreases and its ambiguity increase. As a result, with passage of time due to the increase of the dimensions of the particle and growing of its space, the ambiguity in particle and its space will increase.

The space and time of our humans are so much similar to each other and as a result we can take them as one with a small approximation.

If no special even will occur to us, we are left with

$$\frac{\infty}{2} - 14 \text{ billion}$$

Of our space life. After that all the particles will become dilute so much that they will disappear or they will become one with space and with coincidence of the world with its inverse, the world will disappear totally.

But as it was mentioned, collapsing is an exclusive matter for a particle, and the collapse of world is also an exclusive matter for us.

It is because no particle is able to make all the particles have the same time and space as itself.

As a result. Although all the particles of the world will be participating with us in destroying of the world, but due to having exclusive space, still a lot of amount of space will remain.

A BRIEF EXPLANATION ABOUT LIGHT

As it was described in previous section, light or in general all the electromagnetic waves have totally different nature as to photon. The photon we know is the non-existent photon and due to not having time and dimensions, in fact it doesn't exist.

What we observe from photon is only the C speed. Light has the nature of a wave and since it is riding on photon, it also has the speed of photon. What makes us to see light with such a speed, is time.

Now this question occurs to us that how come a particle which doesn't exists, have movement?

In response we should say that photon always feel itself in inertia status due to not having any time and in fact this is the particle which is moving and is getting far from non-existence in all directions with speed of C and this is turn is because of the penetration of existence in to it.

Light is a wave which is created in the movement of the particle.

If we want to be more exact, we can say if pure energy will

be applied to the movement of the particle, since it is not able change the movement of particle comparing to C, as a result it is seen in the form of a wave in the movement of the particle or light or in general in the form of a electromagnetic waves.

Recently in some of the articles we have seen that it is possible that the speed of a particle to more than C. Regarding this we should say that this is completely untrue. As it was stated previously, movement is against the existence and the particle which reached the speed of C, the photon is in inertia status for it and as a result the time and space will disappear for it. In other worlds only a particle will reach this speed that there is no existence inside of it and it is photon. Therefore, the photon itself cannot get a speed higher than the speed of itself.

GRAVITY OF NEUTRAL ZERO

As it was mentioned in the section of balance mathematics, neutral zero has a virtual gravity and absorbs numbers after zero and before infinite. In reality also this is true. Neutral zero which is the end of the world for particle, will absorb the particle to itself. This gravity is applied to particle by the means of time.

In effect of this gravity the particle and its space is become more expanded gradually and more diluted and at the time of $\frac{\infty}{2}$ will reach the end of its path and its density would become equal to the inverse density and the particle become coincident on its inverse. As a result it will neutralize with its inverse and the particle becomes non-existent and destroyed.

The neutral zero gravity is the contrary of the gravity among the particles. It is because the gravity among particles would decrease te existence coefficient and increases the neutral zero gravity of the existence coefficient of the particle.

Neutral zero gravity is the big bang factor and is applied on all the particles in the world and as much as the existence coefficient of the particle is higher, this gravity will be more on it. It is because the speed of time for that particle is more. Although the gravity between the particles is capable of collapsing some of the part of the world (like collapse in the center of black hole) but it is not capable to synchronize the space of all particles for this purpose or in other and more simple words it cannot make all the particles of the world to collapse in their space and the reason is in the properties of this gravity.

For example if we assume that a black hole has reach a stage that photon is not capable of escaping the center of it. It means that it has lost its center under the effect of gravity. In this case, the black hole is not capable of growth and is not able to increase its mass.

It is because due to downfall of an object inside it, equal to the mass of in center of black hole will change into photon or equal of the mass will be destroyed.

As a result we can say that the black hole is saturated. This is maximum mass it can contain.

This amount of gravity is not able to affect all the particles space and make them to collapse. If we assume that the gravity of all the particles would make their space to collapse, due to the presence of neutral zero gravity, it is impossible, which is because of the non-synchronization of the particles. Wile neutral zero gravity affects all the particles by the means of time and pull them to itself.

Although this gravity is unknown so far but the factor of big bang and the reason of it continuing it is this very gravity.

Although world has a starting and ending point in the space of each particle and it is possible some of the particles will get destroyed to the collapse of the world in their space and some other will get destroyed due to reaching neutral zero but due to existence of the spaces from zero to $\frac{\infty}{2}$ the world will always continue.

This question may occur to us that what is the difference if a particle is destroyed by collapsing or with neutral zero? In response we should say that in collapse, the particle will pull the space edge to itself or big bang toward itself and becomes on with it and in the condition of neutral zero goes

Toward the space edge or big bang and becomes one with it.

In the end we should say that although real zero is neutral zero and due to condition we are living in, this zero is unknown to us. But since both of these zero doesn't exist in this world, in the non-existence world the neutral zero can transform into infinite or certain zero and this is the reason of the existence of real world, because the real world is the distance of certain zero and infinite and neutral zero.

www.ingramcontent.com/pod-product-compliance
Lightning Source LLC
Chambersburg PA
CBHW071126210326
41519CB00020B/6439